John Bredakis

A fascinating look at higher Mathematics (expanded version)

A serious attempt to make my method and the gamma function accessible to anyone dealing with integral calculus

GRIN Verlag

Bibliografische Information der Deutschen Nationalbibliothek:

Die Deutsche Bibliothek verzeichnet diese Publikation in der Deutschen National-
bibliografie; detaillierte bibliografische Daten sind im Internet über http://dnb.d-
nb.de/ abrufbar.

Dieses Werk sowie alle darin enthaltenen einzelnen Beiträge und Abbildungen
sind urheberrechtlich geschützt. Jede Verwertung, die nicht ausdrücklich vom
Urheberrechtsschutz zugelassen ist, bedarf der vorherigen Zustimmung des Verla-
ges. Das gilt insbesondere für Vervielfältigungen, Bearbeitungen, Übersetzungen,
Mikroverfilmungen, Auswertungen durch Datenbanken und für die Einspeicherung
und Verarbeitung in elektronische Systeme. Alle Rechte, auch die des auszugsweisen
Nachdrucks, der fotomechanischen Wiedergabe (einschließlich Mikrokopie) sowie
der Auswertung durch Datenbanken oder ähnliche Einrichtungen, vorbehalten.

Imprint:

Copyright © 2011 GRIN Verlag GmbH
Druck und Bindung: Books on Demand GmbH, Norderstedt Germany
ISBN: 978-3-640-89610-3

GRIN - Your knowledge has value

Der GRIN Verlag publiziert seit 1998 wissenschaftliche Arbeiten von Studenten, Hochschullehrern und anderen Akademikern als eBook und gedrucktes Buch. Die Verlagswebsite www.grin.com ist die ideale Plattform zur Veröffentlichung von Hausarbeiten, Abschlussarbeiten, wissenschaftlichen Aufsätzen, Dissertationen und Fachbüchern.

Visit us on the internet:

http://www.grin.com/

http://www.facebook.com/grincom

http://www.twitter.com/grin_com

The bing bang derivation of **innumerous** mathematical formulas
starting from elementary trigonometry

At the beginning is the Integral Ip

$$p = (n+x)$$

$$Ip = \int e^{r.t} . t^p . dt = \left[e^{r.t} . \left[\sum_{k=0}^{k} (-1)^k . \frac{p!}{[p-k]!} . t^{p-k} . r^{-(k+1)} \right] \right.$$

Selected m\<n

$$\left. + (-1)^{m+1} . r^{-(m+1)} . \frac{p!}{[p-(m+1)]!} . Ip-(m+1) \right]$$

$$\frac{p!}{[p-k]!} = p.(p-1).(p-2). \ldots .[p-(k-1)]$$ **k terms** $k=(1,2,3,..,n,n+1)$ $k>/1$ $\quad$ $Ip-1 = \int e^{r.t} . t^{p-1} .dt$

Choices of r , p , x and intervals of integration

I. $\qquad\qquad\qquad p=(n+x)$

$$p.(p-1).(p-2).(p-3)..(p-n) = x.(x+1).(x+2).(x+3)..(x+n)$$

$$r=-1$$

$$\Gamma(n+x+1) = \left[\prod_{k=0}^{n} (x+k) \right] . \Gamma(x) \quad \int_{0}^{+\infty} e^{-t} . t^{n+x} .dt = \left[\prod_{k=0}^{n} (x+k) \right] . \int_{0}^{+\infty} e^{-t} . t^{x-1} .dt$$

Π stands for products | **Basic property of the gamma function**

II. $\qquad\qquad$ **p=any non negative integer** $\qquad$ John Bredakis
$\qquad\qquad\qquad$ n= (0,1,2,3,...n) $\qquad\qquad\qquad$ method

---

$$In = \int e^{r.x} . x^n .dx = e^{r.x} . \left[\sum_{k=0}^{n} (-1)^k . \frac{n!}{[n-k]!} . x^{n-k} . r^{-(k+1)} \right] + C$$

Successive derivatives , in terms of x , with alternating sign

---

Special case: $\quad e^{r.x} = e^{a.x} . [\cos(b.x) + i.\sin(b.x)]$ $\qquad r=(a+i.b)$

$$In = e^{r.x} . \left[\sum_{k=0}^{n} (-1)^k . \frac{n!}{[n-k]!} . x^{n-k} . \frac{(a-i.b)^{k+1}}{(a^2 + b^2)^{-(k+1)}} \right] + C$$

$$\frac{}{[pk+1 - i.qk+1]}$$

Transferring the i in the nominator

$$----=In---- \qquad -------=Inc---------- \qquad -------=Ins----------$$

$$\int e^{r.x} . x^n .dx = \int e^{a.x} . x^n .\cos(b.x).dx + i.\int e^{a.x} . x^n .\sin(b.x).dx$$

Inc is the real part of **In**
Ins is the imaginary part of **In** , without the **i** in front

III.(a)

$$\Gamma(x) = \int_{0}^{+\infty} e^{-t} \cdot t^{x-1} \cdot dt = \lim_{n \to +\infty} \frac{n! \cdot n^{x}}{x \cdot (x+1) \cdot (x+2) \cdot \ldots \cdot (x+n)} \quad \begin{array}{l} x \#0 \\ x \#-1 \\ x \#-2 \\ \text{etc} \end{array}$$

III.(b)

$$B(x,y) = \frac{\Gamma(x) \cdot \Gamma(y)}{\Gamma(x+y)} \quad \boxed{x,y > 0}$$

$B(x,y)$

$$= \int_{0}^{1} (1-t)^{y-1} \cdot t^{x-1} \cdot dt = \int_{0}^{+\infty} \frac{u^{x-1}}{(1+u)^{x+y}} \cdot du = 2 \cdot \int_{0}^{\pi/2} \sin^{2x-1}\theta \cdot \cos^{2y-1}\theta \cdot d\theta$$

III.(c)   **The solutions of Bessel's equation**

$$x \cdot \frac{d}{dx}\left[x \cdot \frac{dy}{dx} \right] + (b^{2} \cdot x^{2} - p^{2}) \cdot y = 0 \quad \boxed{Or} \quad x^{2} \cdot y'' + x \cdot y' + (b^{2} \cdot x^{2} - p^{2}) \cdot y = 0$$

$0 \backslash < x < +\infty$ p=any positive real number   b=constant#0
Or p=n (n=a positive integer or zero)

- Looking at the graphs of those solutions is like facing -
a different kind of trigonometry

| Fourier's series and integral for classical trigonometry | ─Bessel─────── Expansion in orthogonal functions Complete set of those functions |

And for both the Dirac's heretic isosceles triangle is applied

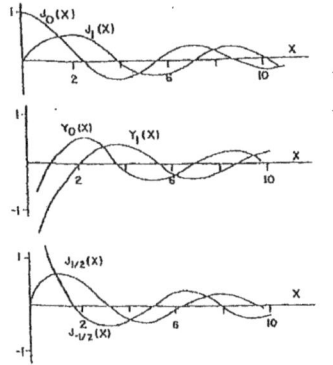

$$Jn(b.x) = \sum_{k=0}^{+\infty} \frac{(-1)^{k}}{k!} \cdot \frac{\left[\dfrac{b.x}{2}\right]^{n+2k}}{\Gamma(k+n+1)}$$

To the memory of my uncle Fotis

J.K.Bredakis MD

The continuity of mathematics: From the integral **Ip** to **innumerous** mathematical formulas , starting from elementary trigonometry.

$$
\begin{array}{l}
p=(n+x) \\[2pt]
Ip = \displaystyle\int e^{r.t}.t^{p}.dt =
\begin{bmatrix}
e^{r.t}.\left[\displaystyle\sum_{k=0}^{m} (-1)^{k} . \dfrac{p!}{[p-k]!}.t^{p-k}.r^{-(k+1)} \right] \\[14pt]
+ (-1)^{m+1}.r^{-(m+1)} . \dfrac{p!}{[p-(m+1)]!}.Ip-(m+1)
\end{bmatrix}\\[6pt]
\text{Selected } m\backslash<n
\end{array}
$$

Choices of r , p , x **and intervals of integration**

$$
\dfrac{p!}{[p-k]!} = p.(p-1).(p-2). \ldots .[p-(k-1)] \qquad \begin{array}{c} \textbf{k terms} \\ k=(1,2,3,..,n,n+1) \quad k>/1 \end{array} \qquad \Bigg| \qquad Ip-1 = \int e^{r.t}.t^{p-1}.dt
$$

Special case: $\quad e^{r.x} = e^{a.x} . [\cos(b.x) + i.\sin(b.x)] \qquad\qquad r=(a+i.b)$

$$
Ip = e^{r.t}.\left[t^{p}.r^{-1} \right] + (-1)^{1}.r^{-1}.p.Ip-1 \hfill m=0
$$

$$
= e^{r.t}.\left[t^{p}.r^{-1} - p.t^{p-1}.r^{-2} \right] + (-1)^{2}.r^{-2}.p.(p-1).Ip-2 \hfill m=1
$$

$$
= e^{r.t}.\left[t^{p}.r^{-1} - p.t^{p-1}.r^{-2} + p.(p-1).t^{p-2}.r^{-3} \right] \hfill m=2
$$
$$
+ (-1)^{3}.r^{-3}.p.(p-1).(p-2).Ip-3
$$

$$
= e^{r.t}.\left[t^{p}.r^{-1} - p.t^{p-1}.r^{-2} + p.(p-1).t^{p-2}.r^{-3} - p.(p-1).(p-2).t^{p-3}.r^{-4} \right] \hfill m=3
$$
$$
+ (-1)^{4}.r^{-4}.p.(p-1).(p-2).(p-3).Ip-4
$$

$$
= e^{r.t}.\left[t^{p}.r^{-1} - p.t^{p-1}.r^{-2} + p.(p-1).t^{p-2}.r^{-3} - p.(p-1).(p-2).t^{p-3}.r^{-4} + p.(p-1).(p-2).(p-3).t^{p-4}.r^{-5} \right] \hfill m=4
$$

$$
\boxed{\text{etc}} \qquad + (-1)^{5}.r^{-5}.p.(p-1).(p-2).(p-3).(p-4).Ip-5
$$

- And then philosophy starts -

The world is not always real as we see it , but has many other
dimensions expressed by the imaginary i.

And this world has strong connection with geometry and trigonometry (converting geometry into algebra) , verifying what the
ancient Greeks said:

    *   - ΑΕΙ   Ο   ΘΕΟΣ   Ο   ΜΕΓΑΣ   ΓΕΩΜΕΤΡΕΙ -
                 3 , 1   4   1   5   9
               The number π by Archimedes

    * The great God always applies the rules of geometry

John.K.Bredakis MD

Assistant Professor University of Athens
American Board Certified Cardiologist

-          Born in Athens Greece 28/11/1946

-        Graduate of the medical school (1970)
               University of Athens

-     Trained in internal medicine and Cardiology
                  (1970-1977)
                 Chicago - USA

- Consultant Cardiologist - Areteion Hospital Athens Greece
                      Since 1977

Thanks God , uncle Fotis , Areteion Hospital
my parents , my wife Sofia
and professors C.Tountas , D.Voros , G.Limouris

Athens Greece 2011

John Bredakis method and Highways without speed limit

The only method providing cook book formulas for a variety of integrals , for any non negative integer n and the relevant improper forms.

1. $\int e^{a.x} . x^n .dx$ $\int \ln^n (a.x) .dx$ 1. (a≠0)

2. $\int e^{a.x} . x^n .\cos(b.x) .dx$ $\int e^{a.x} . x^n .\sin(b.x) .dx$

3. $\int e^{a.x} . x^n .\cosh(b.x) .dx$ $\int e^{a.x} . x^n .\sinh(b.x) .dx$

4. $\int x^n .\cos(b.x) .dx$ $\int x^n .\sin(b.x) .dx$

5. $\int x^n .\cosh(b.x) .dx$ $\int x^n .\sinh(b.x) .dx$

The improper forms of this method are related to the gamma function $\Gamma(x)$ and to the Laplace Transformation

The general formula for In
(n=0,1,2,3,..,n)

$$In = \int e^{r.x} . x^n .dx = e^{r.x} . \left[\sum_{k=0}^{n} (-1)^k . \frac{n!}{[n-k]!} . x^{n-k} . r^{-(k+1)} \right] + C$$

Successive derivatives , in terms of x , with alternating sign

$$Iln = \int \ln^n x.dx = x. \left[\sum_{k=0}^{n} (-1)^k . \frac{n!}{[n-k]!} . \ln^{n-k} x \right] + C$$

Something like successive derivatives with alternating sign

$$Inc = \int e^{a.x} . x^n .\cos(b.x) .dx = e^{a.x} . [..] + C$$

$$\left[\sum_{k=0}^{n} (-1)^k . \frac{n!}{[n-k]!} . x^{n-k} . [p_{k+1}.\cos(b.x) + q_{k+1}.\sin(b.x)] .D^{-(k+1)} \right]$$
$$D=(a^2 + b^2)$$

$$Ins = \int e^{a.x} . x^n .\sin(b.x) .dx = e^{a.x} . [..] + C$$

$$\left[\sum_{k=0}^{n} (-1)^k . \frac{n!}{[n-k]!} . x^{n-k} . [p_{k+1}.\sin(b.x) - q_{k+1}.\cos(b.x)] .D^{-(k+1)} \right]$$
$$D=(a^2 + b^2)$$

Even powers of b	Odd powers of b
$p1=a$	$q1=b$
$p2=(a^2 - b^2)$	$q2=2.a.b$
$p3=(a^3 - 3.a.b^2)$	$q3=(3.a^2.b - b^3)$
$p4=(a^4 - 6.a^2.b^2 + b^4)$	$q4=(4.a^3.b - 4.a.b^3)$
$p5=(a^5 - 10.a^3.b^2 + 5.a.b^4)$	$q5=(5.a^4.b - 10.a^2.b^3 + b^5)$
$p6=(a^6 - 15.a^4.b^2 + 15.a^2.b^4 - b^6)$	$q6=(6.a^5.b - 20.a^3.b^3 + 6.a.b^5)$

etc

The first term is positive and the others alternate in sign

The general formula for Inhc and Inhs
(n=0,1,2,3,..,n)

$$\text{Inhc} = \int e^{a.x}.x^n.\cosh(b.x).dx = e^{a.x}.[..] + C$$

$$\left[\sum_{k=0}^{n} (-1)^k.\frac{n!}{[n-k]!}.x^{n-k}.[pk+1.\cosh(b.x)-qk+1.\sinh(b.x)].D^{-(k+1)} \right]$$
$$D=(a^2 - b^2)$$

$$\text{Inhs} = \int e^{a.x}.x^n.\sinh(b.x).dx = e^{a.x}.[..] + C$$

$$\left[\sum_{k=0}^{n} (-1)^k.\frac{n!}{[n-k]!}.x^{n-k}.[pk+1.\sinh(b.x)-qk+1.\cosh(b.x)].D^{-(k+1)} \right]$$
$$D=(a^2 - b^2)$$

Even powers of b	Odd powers of b
$p1=a$	$q1=b$
$p2=(a^2 + b^2)$	$q2=2.a.b$
$p3=(a^3 + 3.a.b^2)$	$q3=(3.a^2.b + b^3)$
$p4=(a^4 + 6.a^2.b^2 + b^4)$	$q4=(4.a^3.b + 4.a.b^3)$
$p5=(a^5 + 10.a^3.b^2 + 5.a.b^4)$	$q5=(5.a^4.b + 10.a^2.b^3 + b^5)$
$p6=(a^6 + 15.a^4.b^2 + 15.a^2.b^4 + b^6)$	$q6=(6.a^5.b + 20.a^3.b^3 + 6.a.b^5)$

etc

All terms are positive

- Cook book formulas for the following integrals -

A.
$$\int e^x \cdot x^1 \cdot dx = e^x \cdot [x-1] \qquad\qquad\qquad + C \qquad 1=1!$$
$$\int e^x \cdot x^2 \cdot dx = e^x \cdot [x^2 - 2.x + 2] \qquad\qquad + C \qquad 2=2!$$
$$\int e^x \cdot x^3 \cdot dx = e^x \cdot [x^3 - 3.x^2 + 6.x - 6] \qquad + C \qquad 6=3!$$
$$\int e^x \cdot x^4 \cdot dx = e^x \cdot [x^4 - 4.x^3 + 12.x^2 - 24.x + 24] \quad + C \qquad 24=4!$$
etc

Successive derivatives with alternating sign
multiplied by e^x

B.
$$\int e^{-x} \cdot x^1 \cdot dx = - e^{-x} \cdot [x+1] \qquad\qquad\qquad + C \qquad 1=1!$$
$$\int e^{-x} \cdot x^2 \cdot dx = - e^{-x} \cdot [x^2 + 2.x + 2] \qquad\qquad + C \qquad 2=2!$$
$$\int e^{-x} \cdot x^3 \cdot dx = - e^{-x} \cdot [x^3 + 3.x^2 + 6.x + 6] \qquad + C \qquad 6=3!$$
$$\int e^{-x} \cdot x^4 \cdot dx = - e^{-x} \cdot [x^4 + 4.x^3 + 12.x^2 + 24.x + 24] \quad + C \qquad 24=4!$$
etc

Successive derivatives with positive sign
multiplied by $-e^{-x}$

C.
$$\int \ln^1 x.dx = x.[\ln x - 1] \qquad\qquad\qquad\qquad\qquad + C \qquad 1=1!$$
$$\int \ln^2 x.dx = x.[\ln^2 x - 2.\ln x + 2] \qquad\qquad\qquad + C \qquad 2=2!$$
$$\int \ln^3 x.dx = x.[\ln^3 x - 3.\ln^2 x + 6.\ln x - 6] \qquad\quad + C \qquad 6=3!$$
$$\int \ln^4 x.dx = x.[\ln^4 x - 4.\ln^3 x + 12.\ln^2 x - 24.\ln x + 24] \quad + C \qquad 24=4!$$
etc

Something like successive derivatives with alternating sign
multiplied by x

D.
$$\int L^1 .dx = x.[L + 1] \qquad\qquad\qquad\qquad L=\ln(\tfrac{1}{t})=\ln(t^{-1})=-\ln t \quad + C$$
$$\int L^2 .dx = x.[L^2 + 2.L + 2] \qquad\qquad\qquad 2=2! \qquad\qquad\qquad\qquad + C$$
$$\int L^3 .dx = x.[L^3 + 3.L^2 + 6.L + 6] \qquad\qquad 6=3! \qquad\qquad\qquad\qquad + C$$
$$\int L^4 .dx = x.[L^4 + 4.L^3 + 12.L^2 + 24.L + 24] \qquad 24=4! \qquad\qquad\qquad + C$$
etc **All terms positive**

Specific formulas

$$Ioc = \int e^{a.x}.\cos(b.x).dx \qquad\qquad Ios = \int e^{a.x}.\sin(b.x).dx$$

$$e^{a.x}.D^{-1}.[p1.c + q1.s] + C \qquad\qquad e^{a.x}.D^{-1}.[p1.s - q1.c] \,] + C$$

$$D=(a^2 + b^2) \qquad c=\cos(b.x) \qquad s=\sin(b.x) \qquad x^o =1$$

$$I1c = \int e^{a.x}.x^1.\cos(b.x).dx \qquad\qquad I1s = \int e^{a.x}.x^1.\sin(b.x).dx$$

$$e^{a.x}.\begin{bmatrix} x^1.D^{-1}.[p1.c + q1.s] \\ - 1.x^o.D^{-2}.[p2.c + q2.s] \end{bmatrix} + C \qquad e^{a.x}.\begin{bmatrix} x^1.D^{-1}.[p1.s - q1.c] \\ - 1.x^o.D^{-2}.[p2.s - q2.c] \end{bmatrix} + C$$

$$I2c= e^{a.x}.\begin{bmatrix} x^2.D^{-1}.[p1.c + q1.s] \\ - 2.x^1.D^{-2}.[p2.c + q2.s] \\ + 2.x^o.D^{-3}.[p3.c + q3.s] \end{bmatrix} + C \qquad I2s= e^{a.x}.\begin{bmatrix} x^2.D^{-1}.[p1.s - q1.c] \\ - 2.x^1.D^{-2}.[p2.s - q2.c] \\ + 2.x^o.D^{-3}.[p3.s - q3.c] \end{bmatrix} + C$$

$$I3c= e^{a.x}.\begin{bmatrix} x^3.D^{-1}.[p1.c + q1.s] \\ - 3.x^2.D^{-2}.[p2.c + q2.s] \\ + 6.x^1.D^{-3}.[p3.c + q3.s] \\ - 6.x^o.D^{-4}.[p4.c + q4.s] \end{bmatrix} + C \qquad I3s= e^{a.x}.\begin{bmatrix} x^3.D^{-1}.[p1.s - q1.c] \\ - 3.x^2.D^{-2}.[p2.s - q2.c] \\ + 6.x^1.D^{-3}.[p3.s - q3.c] \\ - 6.x^o.D^{-4}.[p4.s - q4.c] \end{bmatrix} + C$$

etc

Even powers of b	**Odd powers of b**
$p1=a$	$q1=b$
$p2=(a^2 - b^2)$	$q2=2.a.b$
$p3=(a^3 - 3.a.b^2)$	$q3=(3.a^2.b - b^3)$
$p4=(a^4 - 6.a^2.b^2 + b^4)$	$q4=(4.a^3.b - 4.a.b^3)$

etc
The first term is positive and the others alternate in sign

$I_{0c} = \int \cos(k.t).dt$	$I_{0s} = \int \sin(k.t).dt$
$[\ +\ 1.k^{-1}.s\ ]\ +\ C$	$[\ -\ 1.k^{-1}.c\ ]\ +\ C$

$$c=\cos(k.t) \qquad s=\sin(k.t) \qquad t^{o}=1$$

$I_{1c} = \int t^{1}.\cos(k.t).dt$	$I_{1s} = \int t^{1}.\sin(k.t).dt$
$\left[\begin{array}{l} +\ 1.t^{1}.k^{-1}.s \\ +\ 1.t^{o}.k^{-2}.c \end{array}\right] + C$	$\left[\begin{array}{l} -\ 1.t^{1}.k^{-1}.c \\ +\ 1.t^{o}.k^{-2}.s \end{array}\right] + C$

$I_{2c} = \int t^{2}.\cos(k.t).dt$	$I_{2s} = \int t^{2}.\sin(k.t).dt$
$\left[\begin{array}{l} +\ 1.t^{2}.k^{-1}.s \\ +\ 2.t^{1}.k^{-2}.c \\ -\ 2.t^{o}.k^{-3}.s \end{array}\right] + C$	$\left[\begin{array}{l} -\ 1.t^{2}.k^{-1}.c \\ +\ 2.t^{1}.k^{-2}.s \\ +\ 2.t^{o}.k^{-3}.c \end{array}\right] + C$

$I_{3c}=$	$I_{3s}=$
$\left[\begin{array}{l} +\ 1.t^{3}.k^{-1}.s \\ +\ 3.t^{2}.k^{-2}.c \\ -\ 6.t^{1}.k^{-3}.s \\ -\ 6.t^{o}.k^{-4}.c \end{array}\right] + C$	$\left[\begin{array}{l} -\ 1.t^{3}.k^{-1}.c \\ +\ 3.t^{2}.k^{-2}.s \\ +\ 6.t^{1}.k^{-3}.c \\ -\ 6.t^{o}.k^{-4}.s \end{array}\right] + C$

etc

.The Inc starts with two positive terms, followed by two negative terms and two positive terms etc (**space permitted**).

.The Ins starts with one negative term, followed by two positive terms and two negative terms etc (**space permitted**).

$$Iohc = \int e^{a.x} .\cosh(b.x) .dx \qquad\bigg|\qquad Iohs = \int e^{a.x} .\sinh(b.x) .dx$$

$$e^{a.x} .D^{-1} .[p1.c - q1.s] + C \qquad\bigg|\qquad e^{a.x} .D^{-1} .[p1.s - q1.c] + C$$

$$D = (a^2 - b^2) \qquad c = \cosh(b.x) \qquad s = \sinh(b.x) \qquad x^o = 1$$

$$I1hc = \int e^{a.x} .x^1 .\cosh(b.x) .dx \qquad\bigg|\qquad I1hs = \int e^{a.x} .x^1 .\sinh(b.x) .dx$$

$$e^{a.x} . \left[\begin{array}{l} x^1 .D^{-1} .[p1.c - q1.s] \\ - 1.x^o .D^{-2} .[p2.c - q2.s] \end{array} \right] + C \qquad\bigg|\qquad e^{a.x} . \left[\begin{array}{l} x^1 .D^{-1} .[p1.s - q1.c] \\ - 1.x^o .D^{-2} .[p2.s - q2.c] \end{array} \right] + C$$

$$I2hc = \int e^{a.x} .x^2 .\cosh(b.x) .dx \qquad\bigg|\qquad I2hs = \int e^{a.x} .x^2 .\sinh(b.x) .dx$$

$$e^{a.x} . \left[\begin{array}{l} x^2 .D^{-1} .[p1.c - q1.s] \\ - 2.x^1 .D^{-2} .[p2.c - q2.s] \\ + 2.x^o .D^{-3} .[p3.c - q3.s] \end{array} \right] + C \qquad\bigg|\qquad e^{a.x} . \left[\begin{array}{l} x^2 .D^{-1} .[p1.s - q1.c] \\ - 2.x^1 .D^{-2} .[p2.s - q2.c] \\ + 2.x^o .D^{-3} .[p3.s - q3.c] \end{array} \right] + C$$

$$I3hc = e^{a.x} . \left[\begin{array}{l} x^3 .D^{-1} .[p1.c - q1.s] \\ - 3.x^2 .D^{-2} .[p2.c - q2.s] \\ + 6.x^1 .D^{-3} .[p3.c - q3.s] \\ - 6.x^o .D^{-4} .[p4.c - q4.s] \end{array} \right] + C \qquad\bigg|\qquad I3hs = e^{a.x} . \left[\begin{array}{l} x^3 .D^{-1} .[p1.s - q1.c] \\ - 3.x^2 .D^{-2} .[p2.s - q2.c] \\ + 6.x^1 .D^{-3} .[p3.s - q3.c] \\ - 6.x^o .D^{-4} .[p4.s - q4.c] \end{array} \right] + C$$

etc

p1=a **Even powers of b**	q1=b **Odd powers of b**
$p2 = (a^2 + b^2)$	$q2 = 2.a.b$
$p3 = (a^3 + 3.a.b^2)$	$q3 = (3.a^2.b + b^3)$

etc
All terms are positive

$Iohc = \int \cosh(k.t).dt$	$Iohs = \int \sinh(k.t).dt$
$k^{-1}.s + C$	$k^{-1}.c + C$

$$c = \cosh(k.t) \qquad s = \sinh(k.t) \qquad t^o = 1$$

$I1hc = \int t^1.\cosh(k.t).dt$	$I1hs = \int t^1.\sinh(k.t).dt$
$\left[\begin{array}{c} t^1.k^{-1}.s \\ -1.t^o.k^{-2}.c \end{array}\right] + C$	$\left[\begin{array}{c} t^1.k^{-1}.c \\ -1.t^o.k^{-2}.s \end{array}\right] + C$

$I2hc = \int t^2.\cosh(k.t).dt$	$I2hs = \int t^2.\sinh(k.t).dt$
$\left[\begin{array}{c} t^2.k^{-1}.s \\ -2.t^1.k^{-2}.c \\ +2.t^o.k^{-3}.s \end{array}\right] + C$	$\left[\begin{array}{c} t^2.k^{-1}.c \\ -2.t^1.k^{-2}.s \\ +2.t^o.k^{-3}.c \end{array}\right] + C$

$I3hc = \int t^3.\cosh(k.t).dt$	$I3hs = \int t^3.\sinh(k.t).dt$
$\left[\begin{array}{c} t^3.k^{-1}.s \\ -3.t^2.k^{-2}.c \\ +6.t^1.k^{-3}.s \\ -6.t^o.k^{-4}.c \end{array}\right] + C$	$\left[\begin{array}{c} t^3.k^{-1}.c \\ -3.t^2.k^{-2}.s \\ +6.t^1.k^{-3}.c \\ -6.t^o.k^{-4}.s \end{array}\right] + C$

etc

.For both (Inhc & Inhs) after the initial term, successive derivatives (in terms of t) with alternating sign.

Improper integrals

$$\int_{0}^{+\infty} e^{-s.t} .\cosh(k.t).dt = \frac{s}{s^2 - k^2} \qquad s>0$$

$$\int_{0}^{+\infty} e^{-s.t} .\sinh(k.t).dt = \frac{k}{s^2 - k^2} \qquad s>0$$

$$D=(s^2 - k^2)>0 \qquad \begin{array}{l}(-s+k)<0 \quad s>0 \\ (-s-k)<0\end{array}$$

$$\int_{0}^{+\infty} e^{-s.t} .t.\cosh(k.t).dt \qquad\qquad 1!.D^{-2}.(s^2 + k^2)$$

$$\int_{0}^{+\infty} e^{-s.t} .t.\sinh(k.t).dt \qquad\qquad 1!.D^{-2}.(2.s.k)$$

$$\int_{0}^{+\infty} e^{-s.t} .t^2 .\cosh(k.t).dt \qquad\qquad 2!.D^{-3}.p3$$

$$\int_{0}^{+\infty} e^{-s.t} .t^2 .\sinh(k.t).dt \qquad\qquad 2!.D^{-3}.q3$$

$$\int_{0}^{+\infty} e^{-s.t} .t^3 .\cosh(k.t).dt \qquad\qquad 3!.D^{-4}.p4$$

$$\int_{0}^{+\infty} e^{-s.t} .t^3 .\sinh(k.t).dt \qquad\qquad 3!.D^{-4}.q4$$

etc

Even powers of k	Odd powers of k
p1=s	q1=k
p2=$(s^2 + k^2)$	q2=2.s.k
p3=$(s^3 + 3.s.k^2)$	q3=$(3.s^2 .k + k^3)$
p4=$(s^4 + 6.s^2 .k^2 + k^4)$	q4=$(4.s^3 .k + 4.s.k^3)$

etc
All terms positive

Improper integrals

$$\int_{0}^{+\infty} e^{-s.t}.\cos(k.t).dt = \frac{s}{s^2 + k^2} \qquad s>0 \qquad\qquad \int_{0}^{+\infty} e^{-s.t}.\sin(k.t).dt = \frac{k}{s^2 + k^2} \qquad s>0$$

$$D=(s^2 + k^2)$$

$$\int_{0}^{+\infty} e^{-s.t}.t.\cos(k.t).dt$$
$$1!.D^{-2}.(s^2 - k^2)$$

$$\int_{0}^{+\infty} e^{-s.t}.t.\sin(k.t).dt$$
$$1!.D^{-2}.(2.s.k)$$

$$\int_{0}^{+\infty} e^{-s.t}.t^2.\cos(k.t).dt$$
$$2!.D^{-3}.p3$$

$$\int_{0}^{+\infty} e^{-s.t}.t^2.\sin(k.t).dt$$
$$2!.D^{-3}.q3$$

$$\int_{0}^{+\infty} e^{-s.t}.t^3.\cos(k.t).dt$$
$$3!.D^{-4}.p4$$

$$\int_{0}^{+\infty} e^{-s.t}.t^3.\sin(k.t).dt$$
$$3!.D^{-4}.q4$$

etc

Even powers of k	Odd powers of k
p1=s	q1=k
p2=$(s^2 - k^2)$	q2=2.s.k
p3=$(s^3 - 3.s.k^2)$	q3=$(3.s^2.k - k^3)$
p4=$(s^4 - 6.s^2.k^2 + k^4)$	q4=$(4.s^3.k - 4.s.k^3)$
	etc
The first term is positive and the others alternate in sign	

- **Laplace Transformation** and the relation of some transforms -
to the gamma function $\Gamma(x)$ and to the John Bredakis method

---

Laplace Transformation: $\displaystyle\int_0^{+\infty} e^{-f(s).t} .f(t).dt = F(s)$ $f(s)>0$

In the simplest form $f(s)=s$ $s>0$

$f(t)$ is a continuous or piece wise continuous function

- -

1. Given a $f(t)$ to find the corresponding $F(s)$
or 2. Given a $F(s)$ to find the corresponding $f(t)$

$\boxed{L[f(t)] = F(s)}$ and $\boxed{L-1[F(s) = f(t)}$

$L-1[F(s)]=f(t)$ means that to a given $F(s)$ corresponds
one and only $f(t)$ such that $L[f(t)]=F(s)$

The 1st shifting theorem

$L[e^{-a.t}.f(t)] = F(s+a)$	$L[e^{a.t}.f(t)] = F(s-a)$
$\displaystyle\int_0^{+\infty} e^{-s.t} .e^{-a.t} .t^n .dt = \dfrac{\Gamma(n+1)}{(s+a)^{n+1}}$	$\displaystyle\int_0^{+\infty} e^{-s.t} .e^{a.t} .t^n .dt = \dfrac{\Gamma(n+1)}{(s-a)^{n+1}}$
$s>0$ $a>0$	$s>0$ $(s-a)>0$

The 2nd shifting theorem via the unit step function $u_a(t)$

$\displaystyle\int_a^{+\infty} e^{-s.t} .f(t).dt = \displaystyle\int_a^{+\infty} e^{-s.(t+a)} .f(t+a).d(t+a)$ **By independence of notation**

$----L_a[f(t)]----$ $s>0$ $a>0$

$= e^{-a.s} . \displaystyle\int_0^{+\infty} e^{-s.t} .f(t+a).dt = e^{-a.s} .L[f(t+a)]$

- -

$L[u_a(t).f(t)] = L_a[f(t)] = e^{-a.s} .L[f(t+a)] = e^{-a.s} .F(s,a)$ $s>0$ $a>0$

Corollary to the 2nd shifting theorem:

$L[u_a(t).f(t-a)] = L_a[f(t-a)] = e^{-a.s} .L[f(t)] = e^{-a.s} .F(s)$ $s>0$ $a>0$

Instead of providing a table of inverse Laplace Transforms

- Lets focus on the examples below -

$$L^{-1}\left[\frac{n!}{s^{n+1}}\right] = t^n \qquad n! = \Gamma(n+1)$$

$$L^{-1}\left[\frac{n!}{(s-a)^{n+1}}\right] = t^n \cdot e^{a \cdot t}$$

$$\int_0^{+\infty} e^{-a \cdot x} \cdot \sin(b \cdot x) \, dx = \frac{b}{a^2 + b^2} \qquad a > 0$$

$$\int_0^{+\infty} e^{-a \cdot x} \cdot \cos(b \cdot x) \, dx = \frac{a}{a^2 + b^2} \qquad a > 0$$

$$\int_0^{+\infty} x \cdot e^{-a \cdot x} \cdot \sin(b \cdot x) \, dx = \frac{2 \cdot b \cdot a}{(a^2 + b^2)^2} \qquad a > 0$$

$$\int_0^{+\infty} x \cdot e^{-a \cdot x} \cdot \cos(b \cdot x) \, dx = \frac{a^2 - b^2}{(a^2 + b^2)^2} \qquad a > 0$$

$$L^{-1}\left[\frac{b}{s^2 + b^2}\right] = \sin(b \cdot t)$$

$$L^{-1}\left[\frac{s}{s^2 + b^2}\right] = \cos(b \cdot t)$$

$$L^{-1}\left[\frac{b}{(s-a)^2 + b^2}\right] = e^{a \cdot t} \cdot \sin(b \cdot t)$$

$$L^{-1}\left[\frac{(s-a)}{(s-a)^2 + b^2}\right] = e^{a \cdot t} \cdot \cos(b \cdot t)$$

$$L^{-1}\left[\frac{b \cdot s}{\left[s^2 + b^2\right]^2}\right] = \frac{1}{2} \cdot t \cdot \sin(b \cdot t)$$

$$L^{-1}\left[\frac{s^2 - b^2}{\left[s^2 + b^2\right]^2}\right] = t \cdot \cos(b \cdot t)$$

$$L^{-1}\left[\frac{b}{s^2 - b^2}\right] = \sinh(b \cdot t)$$

$$L^{-1}\left[\frac{s}{s^2 - b^2}\right] = \cosh(b \cdot t)$$

A summary of the gamma $\Gamma(x)$ and the beta $B(x,y)$ function

$$\Gamma(x) = \int_0^{+\infty} e^{-t} t^{x-1} dt = \lim_{n \to +\infty} \frac{n! \cdot n^x}{x.(x+1).(x+2). \ldots .(x+n)} \qquad \begin{array}{l} x\#0 \\ x\#-1 \\ x\#-2 \\ etc \end{array}$$

Main property: $x.\Gamma(x) = \Gamma(x+1)$

$$\Gamma(x) = \frac{\Gamma(x+1)}{x} = \frac{\Gamma(x+2)}{x.(x+1)} = \frac{\Gamma(x+3)}{x.(x+1).(x+2)} etc \left| \frac{\Gamma(x)}{\Gamma(n+x+1)} = \frac{1}{x.(x+1) .. (x+n)} \right.$$

$$\Gamma(n+1) = n! \quad \left| \quad \Gamma(n+\tfrac{1}{2}) = \frac{(2n)!}{2^{2n}.n!} . \sqrt{\pi} = \frac{(2n)!}{2^n.n!} . \frac{1}{2^n} . \sqrt{\pi} \right.$$

$$\Gamma(1) = 0! = 1 \qquad \Gamma(2) = 1! = 1 \qquad \Gamma(3) = 2! = 2 \qquad \Gamma(4) = 3! = 6 \qquad \Gamma(5) = 4! = 24 \textbf{ etc}$$

$$\Gamma(\tfrac{1}{2}) = \sqrt{\pi} \left| \Gamma(\tfrac{3}{2}) = \tfrac{1}{2}.\sqrt{\pi} \right| \Gamma(\tfrac{5}{2}) = \tfrac{3}{2}.\tfrac{1}{2}.\sqrt{\pi} \left| \Gamma(\tfrac{7}{2}) = \tfrac{5}{2}.\tfrac{3}{2}.\tfrac{1}{2}.\sqrt{\pi} \right| \Gamma(\tfrac{9}{2}) = \tfrac{7}{2}.\tfrac{5}{2}.\tfrac{3}{2}.\tfrac{1}{2}.\sqrt{\pi} \textbf{ etc}$$

$$\int_0^{+\infty} e^{-s.t} t^{x-1} dt = 2.\int_0^{+\infty} e^{-s.t^2} t^{2x-1} dt = \int_0^1 t^{s-1} . \left[\ln(\tfrac{1}{t}) \right]^{x-1} dt = \frac{\Gamma(x)}{s^x}$$

$$s>0 \; x\#0,-1,-2,etc \qquad s>0 \qquad x>0 \qquad s>0 \quad s\#x \quad x>0$$

Valid also for s=1

$$\begin{array}{l} \textbf{Legendre's} \\ \textbf{dublication formula} \end{array} \qquad \Gamma(2x) = \frac{1}{\sqrt{\pi}} . 2^{2x-1} . \Gamma(x) . \Gamma(x+\tfrac{1}{2})$$

$$x>0$$

$$\Gamma(x).\Gamma(1-x) = \frac{\pi}{\sin(\pi.x)} \left| \Gamma(x).\Gamma(-x) = \frac{\pi}{-x.\sin(\pi.x)} \right| \Gamma(\tfrac{1}{2}+x).\Gamma(\tfrac{1}{2}-x) = \frac{\pi}{\cos(\pi.x)}$$

Denominator # 0

$$\frac{1}{\Gamma(x)} = \lim_{n\to+\infty} \frac{x.(x+1).(x+2)..(x+n)}{n!.n^x} = e^{\gamma.x}.x.\prod_{n=1}^{+\infty}\left[1+\frac{x}{n}\right].e^{-x/n}$$

$\prod$ stands for products

$$\gamma = \lim_{n\to+\infty}\left[1+\frac{1}{2}+\frac{1}{3}+..+\frac{1}{n}-\ln n\right] = 0.5777216 = \text{Euler's constant}$$

From $\ln\left[\dfrac{1}{\Gamma(x)}\right]$ to: $\dfrac{\Gamma'(x)}{\Gamma(x)} = \displaystyle\sum_{n=1}^{+\infty}\left[\dfrac{1}{n}-\dfrac{1}{n+x}\right] - \gamma - \dfrac{1}{x}$

$$\frac{\Gamma'(k+1)}{\Gamma(k+1)} = 1 + \frac{1}{2} + \frac{1}{3} + \frac{1}{4} + \frac{1}{5} + \frac{1}{6} + \frac{1}{7} +..+ \frac{1}{k} - \gamma = \left[\varphi(k) - \gamma\right]$$

k=a positive integer or zero $\qquad \varphi(0)=0$ by convention

$\dfrac{\Gamma'(1)}{\Gamma(1)} = -\gamma$ | $\dfrac{\Gamma'(2)}{\Gamma(2)} = 1-\gamma$ | $\dfrac{\Gamma'(3)}{\Gamma(3)} = 1+\dfrac{1}{2}-\gamma$ | $\dfrac{\Gamma'(4)}{\Gamma(4)} = 1+\dfrac{1}{2}+\dfrac{1}{3}-\gamma$

$$B(x,y) = \frac{\Gamma(x).\Gamma(y)}{\Gamma(x+y)} \quad \boxed{x,y > 0}$$

$B(x,y)$

$$= \int_0^1 (1-t)^{y-1}.t^{x-1}.dt = \int_0^{+\infty}\frac{u^{x-1}}{(1+u)^{x+y}}.du = 2.\int_0^{\pi/2}\sin^{2x-1}\theta.\cos^{2y-1}\theta.d\theta$$

Schematic representation of $\Gamma(x)$

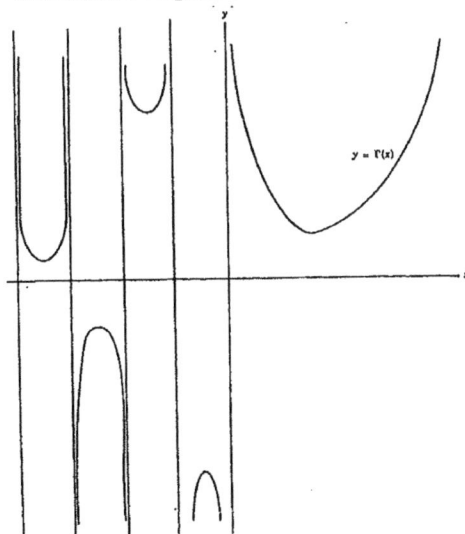

$y = \Gamma(x)$

Numerical table of Γ(x)

x	Γ(x)
1.0	1.00000
.01	0.99433
.02	0.98884
.03	0.98355
.04	0.97844
.05	0.97350
.06	0.96874
.07	0.96415
.08	0.95973
.09	0.95546
1.10	0.95135
.11	0.94740
.12	0.94359
.13	0.93993
.14	0.93642
1.15	0.93304
.16	0.92980
.17	0.92670
.18	0.92373
.19	0.92089
1.20	0.91817
.21	0.91558
.22	0.91311
.23	0.91075
.24	0.90852

x	Γ(x)
1.25	0.90840
.26	0.90440
.27	0.90250
.28	0.90072
.29	0.89904
1.30	0.89747
.31	0.89600
.32	0.89600
.33	0.89464
.34	0.89222
1.35	0.89115
.36	0.89018
.37	0.88931
.38	0.88854
.39	0.88785
1.40	0.88726
.41	0.88676
.42	0.88636
.43	0.88604
.44	0.88581
1.45	0.88566
.46	0.88560
.47	0.88563
.48	0.88575
.49	0.88595

x	Γ(x)
1.50	0.88623
.51	0.88659
.52	0.88704
.53	0.88757
.54	0.88818
1.55	0.88887
.56	0.88964
.57	0.89049
.58	0.89142
.59	0.89243
1.60	0.89352
.61	0.89468
.62	0.89592
.63	0.89724
.64	0.89864
1.65	0.90012
.66	0.90167
.67	0.90330
.68	0.90500
.69	0.90678
1.70	0.90864
.71	0.91057
.72	0.91258
.73	0.91467
.74	0.91683

x	Γ(x)
1.75	0.91906
.76	0.92137
.77	0.92376
.78	0.92623
.79	0.92877
1.80	0.93138
.81	0.93408
.82	0.93685
.83	0.93969
.84	0.94261
1.85	0.94561
.86	0.94869
.87	0.95184
.88	0.95507
.89	0.95838
1.90	0.96177
.91	0.96523
.92	0.96877
.93	0.97240
.94	0.97610
1.95	0.97988
.96	0.98374
.97	0.98768
.98	0.99171
.99	0.99581
2.0	1.00000

Main property: $x.\Gamma(x) = \Gamma(x+1)$

Π stands for products

$$\frac{1}{\Gamma(x)} = \lim_{n \to +\infty} \frac{x.(x+1).(x+2)..(x+n)}{n!.n^x} = \frac{x . \prod_{n=1}^{+\infty}\left[1 + \frac{x}{n}\right]}{n^x}$$

$$x \neq 0, -1, -2, -3, -4 \text{ etc} \qquad = e^{\gamma.x} . x . \prod_{n=1}^{+\infty}\left[1 + \frac{x}{n}\right].e^{-x/n}$$

$$\gamma = \lim_{n \to +\infty}\left[1 + \frac{1}{2} + \frac{1}{3} + .. + \frac{1}{n} - \ln n\right] = 0.5777216 = \text{Euler's constant}$$

Contribution of the gamma function $\Gamma(x)$ to the normal distribution

$\int_0^{+\infty} e^{-s.t^2}.dt = \dfrac{\Gamma(1/2)}{2.s^{1/2}}$ $s>0$	$\int_0^{+\infty} e^{-s.t^2}.t.dt = \dfrac{\Gamma(1)}{2.s}$ $s>0$	$\int_0^{+\infty} e^{-s.t^2}.t^2.dt = \dfrac{\Gamma(3/2)}{2.s^{3/2}}$ $s>0$
Even	Odd	Even

Integrand functions

$\int_{-\infty}^{+\infty} \sqrt{\dfrac{s}{\pi}}.e^{-s.t^2}.dt = 1$ $s>0$	$\int_{-\infty}^{+\infty} \sqrt{\dfrac{s}{\pi}}.e^{-s.t^2}.t.dt = 0$ $s>0$	$\int_{-\infty}^{+\infty} \sqrt{\dfrac{s}{\pi}}.e^{-s.t^2}.t^2.dt = \dfrac{1}{2.s}$ $s>0$

Setting instead of $\sqrt{s}$ the $1/[\sigma.\sqrt{2}]$ $\sigma>0$ & $(x-x_o)$ instead of t we get:

Total probability	Mathematical expectation	Variance
$\int_{-\infty}^{+\infty} \varphi(x).dx = 1$	$\int_{-\infty}^{+\infty} \varphi(x).x.dx = x_o$	$\int_{-\infty}^{+\infty} \varphi(x).(x-x_o)^2.dx = \sigma^2$

$\varphi(t) = \dfrac{1}{\sigma.\sqrt{2\pi}}.e^{-\frac{1}{2}.\left(\frac{t}{\sigma}\right)^2}$	$\varphi(x) = \dfrac{1}{\sigma.\sqrt{2\pi}}.e^{-\frac{1}{2}.\left(\frac{x-x_o}{\sigma}\right)^2}$ **By shifting of axis t**

Non Conditional Probability
Random Variable X={x_n} - Random Numbers $\{x_n\}=\{x_1, x_2, .., x_n\}$

$\Phi(x)$=Probability distribution $\varphi(x)$=Probability density	$P(a\backslash<X<b)=\Phi(x)\Big|_a^b = \int_a^b \varphi(x).dx$

- The integral is replaced by the sum for discrete random variables

- The $\Phi(x)$ for continuous random variables is the area under the probability density function $\varphi(x)$.

- We can start with values of X within the range of reality or from $-\infty$ to $+\infty$ considering the $\varphi(x)$ outside the range of reality as zero.

Applications of the beta function $B(x,y)$

$$B(x,y) = \frac{\Gamma(x).\Gamma(y)}{\Gamma(x+y)} \qquad \boxed{x,y > 0}$$

$$B(x,y) = \int_0^1 (1-t)^{y-1}.t^{x-1}.dt = \int_0^{+\infty} \frac{u^{x-1}}{(1+u)^{x+y}}.du = 2.\int_0^{\pi/2} \sin^{2x-1}\theta.\cos^{2y-1}\theta.d\theta$$

$$k = a \text{ positive integer} - \boxed{x = t^k} \Rightarrow \boxed{t^z = x} \quad z = (1/k)$$

$$z = (1/k)$$

$$\int_0^{+\infty} \frac{dx}{1+x^k} = \int_0^{+\infty} \frac{dt^{1/k}}{1+t} = \frac{1}{k}.\int_0^{+\infty} \frac{t^{z-1}}{1+t}.dt = \frac{1}{k}.\frac{\Gamma(z).\Gamma(1-z)}{\Gamma(1)} = \frac{1}{k}.\frac{\pi}{\sin(\pi.z)}$$

$$z = (1/4)$$

$$\int_0^{+\infty} \frac{dx}{1+x^4} = \frac{1}{4}.\int_0^{+\infty} \frac{t^{z-1}}{1+t}.dt = \frac{1}{4}.\frac{\Gamma(1/4).\Gamma(3/4)}{\Gamma(1)} = \frac{1}{4}.\pi.\sqrt{2}$$

$$\int_0^1 \frac{dx}{\sqrt[k]{1-x}} = \frac{1}{k}.\int_0^1 (1-t)^{-(1/2)}.t^{z-1}.dt = \frac{1}{k}.\frac{\Gamma(1/2).\Gamma(z)}{\Gamma(\frac{1}{k} + \frac{1}{2})} \qquad \boxed{z = (1/k)}$$

$$\int_0^1 \frac{dx}{\sqrt[3]{1-x}} = \frac{1}{3}.\int_0^1 (1-t)^{-(1/2)}.t^{z-1}.dt = \frac{1}{3}.\frac{\Gamma(1/2).\Gamma(1/3)}{\Gamma(\frac{1}{3} + \frac{1}{2})} \qquad \boxed{z = (1/3)}$$

$$\int_0^1 t^m.(1-t)^n.dt = B(m+1,n+1) = \frac{\Gamma(m+1).\Gamma(n+1)}{\Gamma(m+n+2)} = \frac{m!.n!}{(m+n+1)!}$$

$$\int_0^\pi \sin^{2x-1}\theta.d\theta = 2.\int_0^{\pi/2} \sin^{2x-1}\theta.d\theta = B(x,\frac{1}{2}) = \frac{\Gamma(x).\Gamma(1/2)}{\Gamma(x+1/2)} \qquad \boxed{x > 0}$$

- From the integral Ip to the formula of the gamma function $\Gamma(x)$ -

$$p = (n+x)$$

$$Ip = \int e^{r.t}.t^{p}.dt = \left[e^{r.t}.\left[\sum_{k=0}^{m} (-1)^{k}.\frac{p!}{[p-k]!}.t^{p-k}.r^{-(k+1)} \right] + (-1)^{m+1}.r^{-(m+1)}.\frac{p!}{[p-(m+1)]!}.Ip-(m+1) \right]$$

Selected m\<n

Choices of r , p , x and intervals of integration

$$\frac{p!}{[p-k]!} = p.(p-1).(p-2).\ \ldots\ .[p-(k-1)] \qquad \text{k terms} \qquad k=(1,2,3,\ldots,n,n+1) \qquad k \geqslant /1 \qquad\qquad Ip-1 = \int e^{r.t}.t^{p-1}.dt$$

p=(n+x) and r=-1

$$p.(p-1).(p-2).(p-3)\ldots(p-n) = x.(x+1).(x+2).(x+3)\ldots(x+n)$$

$$\Gamma(n+x+1) = \left[\prod_{k=0}^{n} (x+k) \right].\Gamma(x) \qquad \int_{0}^{+\infty} e^{-t}.t^{n+x}.dt = \left[\prod_{k=0}^{n} (x+k) \right].\int_{0}^{+\infty} e^{-t}.t^{x-1}.dt$$

Π stands for products $\qquad$ **Basic property of the gamma function**

$$B(x,y) = \int_{0}^{1} (1-t)^{y-1}.t^{x-1}.dt = \frac{\Gamma(x).\Gamma(y)}{\Gamma(x+y)} \qquad \boxed{x,y > 0}$$

$$\int_{0}^{+\infty} e^{-t}.t^{x-1}.dt = \lim_{n \to +\infty} \int_{0}^{+\infty} \left[1 - \frac{t}{n} \right]^{n}.t^{x-1}.dt$$

By independence of notation $\qquad = \lim_{n \to +\infty} \int_{0}^{+\infty} \left[1 - \frac{n.t}{n} \right]^{n}.(n.t)^{x-1}.d(n.t)$

$$\Gamma(n+1) = n!$$

$$\lim_{n \to +\infty} n^{x}.\frac{\Gamma(n+1).\Gamma(x)}{\Gamma(n+x+1)} = \lim_{n \to +\infty} n^{x}.\frac{1}{\int_{0}^{} (1-t)^{n}.t^{x-1}.dt}$$

Therefore: $\lim_{n \to +\infty} \Gamma(n+x+1) = n^{x}.n!$

And:

$$\Gamma(x) = \int_{0}^{+\infty} e^{-t}.t^{x-1}.dt = \lim_{n \to +\infty} \frac{n!.n^{x}}{x.(x+1).(x+2).\ \ldots\ .(x+n)} \qquad \begin{array}{l} x \# 0 \\ x \# -1 \\ x \# -2 \\ \text{etc} \end{array}$$

- Formula of $B(x,y)$ and the Jacobian -

$$\Gamma(x) = \int_0^{+\infty} e^{-u} \cdot u^{x-1} \cdot du \qquad \Gamma(y) = \int_0^{+\infty} e^{-v} \cdot v^{y-1} \cdot dv$$

$$\Gamma(x) \cdot \Gamma(y) = \int_0^{+\infty} \int_0^{+\infty} e^{-(u+v)} \cdot u^{x-1} \cdot v^{y-1} \cdot du \cdot dv$$

Setting $u=p^2$, $v=s^2$ and switching to polar coordinates

$$\Gamma(x) \cdot \Gamma(y) = 4 \cdot \int_0^{+\infty} \cdot \int_0^{+\infty} e^{-(p^2 + s^2)} \cdot p^{2x-1} \cdot s^{2y-1} \cdot dp \cdot ds \qquad \boxed{\begin{array}{l} p = \rho \cdot \cos\theta \\ s = \rho \cdot \sin\theta \end{array}}$$

$$0 \backslash< \rho \ \backslash<+\infty \qquad 0 \backslash< \theta \ \backslash< \pi/2$$

$$= \Gamma(x) \cdot \Gamma(y) = 4 \cdot \int_0^{\pi/2} \cdot \int_0^{+\infty} e^{-\rho^2} \cdot (\rho \cdot \cos\theta)^{2x-1} \cdot (\rho \cdot \sin\theta)^{2y-1} \cdot \rho \cdot d\rho \cdot d\theta$$

$$= \left[2 \cdot \int_0^{+\infty} e^{-\rho^2} \cdot \rho^{2 \cdot (x+y)-1} \cdot d\rho \right] \cdot \left[2 \cdot \int_0^{\pi/2} \cos^{2x-1}\theta \cdot \sin^{2y-1}\theta \cdot d\theta \right]$$

$$\underbrace{}_{\Gamma(x+y)} \qquad \underbrace{}_{B(x,y)}$$

$$= \Gamma(x+y) \cdot B(x,y)$$

$$B(x,y) = \frac{\Gamma(x) \cdot \Gamma(y)}{\Gamma(x+y)} \qquad \boxed{x,y > 0}$$

$B(x,y)$

$$= \int_0^1 (1-t)^{y-1} \cdot t^{x-1} \cdot dt = \int_0^{+\infty} \frac{u^{x-1}}{(1+u)^{x+y}} \cdot du = 2 \cdot \int_0^{\pi/2} \sin^{2x-1}\theta \cdot \cos^{2y-1}\theta \cdot d\theta$$

Formulas and remarks related to the gamma function $\Gamma(x)$

Derived by Wallis product formula	$\sqrt{\pi} = \Gamma(1/2) = \lim\limits_{n \to +\infty} \dfrac{(2^n . n!)^2}{(2n)!} \cdot \dfrac{1}{\sqrt{n}}$

The rules for the factorials

All 2n factorials $= 1.2.3. \ \ldots .(2n-1).2n \qquad = \qquad (2n)!$

1.	Even factorials $= 2.4.6. \ \ldots .(2n-2).2n \qquad = \quad 2^n . n!$
2.	Odd factorials $= 1.3.5. \ \ldots .(2n-3).(2n-1) = \dfrac{(2n)!}{2^n . n!}$

3.	Ratio of odd factorials up to 2n-1 by even factorials up to 2n $= \dfrac{(2n)!}{(2^n . n!)^2}$

4.	Ratio of odd factorials up to 2n-3 by even factorials up to 2n $= \dfrac{(2n)!}{(2^n . n!)^2} \cdot \dfrac{1}{(2n-1)}$

Π stands for products

$$\frac{\sin x}{x} = \prod_{k=1}^{+\infty} \left[1 - \frac{x^2}{\pi^2 . k^2} \right] \qquad \cos x = \prod_{k=0}^{+\infty} \left[1 - \frac{x^2}{\pi^2 . \left[k + \frac{1}{2} \right]^2} \right]$$

$$\frac{\sin(\pi.x)}{\pi.x} = \prod_{k=1}^{+\infty} \left[1 - \frac{x^2}{k^2} \right] \qquad \cos(\pi.x) = \prod_{k=0}^{+\infty} \left[1 - \frac{x^2}{\left[k + \frac{1}{2} \right]^2} \right]$$

Various proofs related to the gamma $\Gamma(x)$ and beta function $B(x,y)$

The proof that:

$$\int_{0}^{+\infty} e^{-t}.t^{x-1}.dt = 2.\int_{0}^{+\infty} e^{-t^2}.t^{2x-1}.dt = \int_{0}^{1} \left[\ln\left(\frac{1}{t}\right)\right]^{x-1}.dt = \Gamma(x)$$
$$x\#0,-1,-2,etc \qquad\qquad x>0 \qquad\qquad\qquad x>0$$

$$\int_{0}^{+\infty} e^{-t}.t^{x-1}.dt = \int_{0}^{+\infty} e^{-t^2}.\left[t^2\right]^{x-1}.dt = 2.\int_{0}^{+\infty} e^{-t^2}.t^{2x-1}.dt$$
$$x\#0,-1,-2,etc \qquad\qquad x>0 \qquad\qquad\qquad x>0$$

$$L=\left[\ln\left(\frac{1}{t}\right)\right]=\ln t^{-1}=-\ln t \qquad \Box \qquad \Gamma(x) = \int_{0}^{+\infty} e^{-t}.t^{x-1}.dt = \int_{0}^{+\infty} e^{-L}.L^{x-1}.dL$$
$$x>0$$

$$\Gamma(x) = -\int_{1}^{0} t.t^{-1}.L^{x-1}.dt = \int_{0}^{1} \left[\ln\left(\frac{1}{t}\right)\right]^{x-1}.dt \qquad \boxed{x>0}$$

By the same way: setting s.t instead of t
$$s>0$$

$$\int_{0}^{+\infty} e^{-t}.t^{x-1}.dt = \int_{0}^{+\infty} e^{-s.t}.\left[s.t\right]^{x-1}.d(s.t) = s^x.\int_{0}^{+\infty} e^{-s.t}.t^{x-1}.dt$$
$$-----=\Gamma(x)---- \qquad\qquad\qquad s>0 \qquad s\#x$$

$$\int_{0}^{1} \left[\ln\left(\frac{1}{s}\right)\right]^{x-1}.dt = \int_{0}^{1} \left[s.-\ln t\right]^{x-1}.dt = s^x.\int_{0}^{1} t^{s-1}.\left[\ln\left(\frac{1}{t}\right)\right]^{x-1}.dt$$
$$------=\Gamma(x)------ \qquad\qquad s>0 \quad s\#x \qquad\qquad x>0$$

Therefore:

$$\int_{0}^{+\infty} e^{-s.t}.t^{x-1}.dt = 2.\int_{0}^{+\infty} e^{-s.t^2}.t^{2x-1}.dt = \int_{0}^{1} t^{s-1}.\left[\ln\left(\frac{1}{t}\right)\right]^{x-1}.dt = \frac{\Gamma(x)}{x^s}$$
$$s>0 \; x\#0,-1,-2,etc \qquad s>0 \quad x>0 \qquad\qquad s>0 \; s\#x \quad x>0 \qquad s$$

$$B(x,y) = \frac{\Gamma(x).\Gamma(y)}{\Gamma(x+y)} \qquad \boxed{x,y > 0}$$

$B(x,y)$

$$= \int_0^1 (1-t)^{y-1}.t^{x-1}.dt = \int_0^{+\infty} \frac{u^{x-1}}{(1+u)^{x+y}}.du = 2.\int_0^{\pi/2} \sin^{2x-1}\theta.\cos^{2y-1}\theta.d\theta$$

$$\int_0^{+\infty} \frac{\xi^{x-1}}{(1+\xi)^{x+y}}.d\xi = 2.\int_0^{+\infty} \frac{(t^2)^{x-1}}{(1+t^2)^{x+y}}.t.dt \qquad \sqrt{1+t^2}=\sec u, \quad t=\tan u$$

$$= 2.\int_0^{\pi/2} \sin^{2x-2}u.\sec^{2x-2}u.\cos^{2x+2y}u.\sin u.\sec u.\sec^2 u.du \qquad {----=t---\ \ ---=dt--}$$

$$= 2.\int_0^{\pi/2} \sin^{2x-1}u.\cos^{2y-1}u.du = \boxed{I}$$

$$\int_0^1 (1-\xi)^{y-1}.\xi^{x-1}.d\xi = 2.\int_0^1 (1-t^2)^{y-1}.(t^2)^{x-1}.t.dt$$

$$= 2.\int_0^1 (\cos^2 u)^{y-1}.(\sin^2 u)^{x-1}.\sin u.d\sin u$$

$$\boxed{I} = 2.\int_0^{\pi/2} (\cos^2 u)^{y-1}.(\sin^2 u)^{x-1}.\sin u.\cos u.du$$

$$\textbf{Legendre's dublication formula} \qquad \Gamma(2x) = \frac{1}{\sqrt{\pi}} \cdot 2^{2x-1} \cdot \Gamma(x) \cdot \Gamma(x+\frac{1}{2}) \qquad x>0$$

The proof:

$$\frac{1}{2^{2x-1}} = \frac{B(x,x)}{B(x,1/2)} = \frac{\Gamma(x) \cdot \Gamma(x)/\Gamma(2x)}{\Gamma(x) \cdot \Gamma(1/2)/\Gamma(x+1/2)} = \frac{\Gamma(x) \cdot \Gamma(x+1/2)}{\Gamma(2x) \cdot \Gamma(1/2)} \qquad \boxed{x>0}$$

$$B(x,x) = 2 \cdot \int_0^{\pi/2} \sin^{2x-1} u \cdot \cos^{2x-1} u \cdot du = 2^{1-2x} \cdot \int_0^{\pi} \sin^{2x-1}(2u) \cdot d(2u)$$

$$\Gamma(1/2) = \sqrt{\pi} \qquad\qquad\qquad\qquad = \frac{1}{2^{2x-1}} \cdot B(x,1/2)$$

$$B(x,1/2) = 2 \cdot \int_0^{\pi/2} \sin^{2x-1} u \cdot \cos^{2x-1}\Big|_{x=1/2} u \cdot du = \int_0^{\pi} \sin^{2x-1} u \cdot du$$

$$\int_0^{\pi} \sin^{2x-1}\theta \cdot d\theta = \int_0^{\pi/2} \sin^{2x-1}\theta \cdot d\theta + \int_{\pi/2}^{\pi} \sin^{2x-1}\theta \cdot d\theta$$

By independence of notation

$$\int_{\pi/2}^{\pi} \sin^{2x-1}\theta \cdot d\theta = \int_{\pi/2}^{\pi} \sin^{2x-1}(\theta + \frac{\pi}{2}) \cdot d(\theta + \frac{\pi}{2})$$

$$= \int_0^{\pi/2} \sin^{2x-1}(\theta + \frac{\pi}{2}) \cdot d\theta$$

$$= \int_0^{\pi/2} \Big[\sin\theta.\cos\frac{\pi}{2} + \cos\theta.\sin\frac{\pi}{2}\Big]^{2x-1} \cdot d\theta$$

$$= \int_0^{\pi/2} \cos^{2x-1}\theta . d\theta = \frac{1}{2}.B(\frac{1}{2},x)$$

The proof that: $\Gamma(x).\Gamma(1-x) = \dfrac{\pi}{\sin(\pi.x)}$ Denominator $\neq 0$

Setting instead of x the x+(1/2)

$\Gamma(\dfrac{1}{2} + x).\Gamma(\dfrac{1}{2} - x) = \dfrac{\pi}{\cos(\pi.x)}$ Denominator $\neq 0$

Π stands for products

$$\dfrac{\sin x}{x} = \prod_{k=1}^{+\infty}\left[1 - \dfrac{x^2}{\pi^2.k^2}\right] \qquad \cos x = \prod_{k=0}^{+\infty}\left[1 - \dfrac{x^2}{\pi^2.\left[k + \dfrac{1}{2}\right]^2}\right]$$

- -

$$\dfrac{\sin(\pi.x)}{\pi.x} = \prod_{k=1}^{+\infty}\left[1 - \dfrac{x^2}{k^2}\right] \qquad \cos(\pi.x) = \prod_{k=0}^{+\infty}\left[1 - \dfrac{x^2}{\left[k + \dfrac{1}{2}\right]^2}\right]$$

$$\Gamma(x) = \lim_{n \to +\infty} \dfrac{n!.n^x}{x.(x+1).\ \ldots\ .(x+n)} \qquad \Gamma(1-x) = \lim_{n \to +\infty} \dfrac{n!.n^{y=(1-x)}}{y.(y+1).\ \ldots\ .(y+n)}$$

$$\Gamma(x).\Gamma(1-x) = \lim_{n \to +\infty} \dfrac{(n!)^2.n}{x.(1 - x^2).(2^2 - x^2).\ \ldots\ .(n^2 - x^2).(n+1-x)}$$

$$\Gamma(x).\Gamma(1-x) = \lim_{n \to +\infty} \dfrac{1}{x.\prod_{k=1}^{+\infty}\left[1 - \dfrac{x^2}{k^2}\right]} = \dfrac{1}{x.\dfrac{\sin(\pi.x)}{\pi.x}} = \dfrac{\pi}{\sin(\pi.x)}$$

The Wallis product formula:

As $n \to +\infty$ $(n+\frac{1}{2})^2 . \Gamma^2(n+\frac{1}{2}) = \Gamma^2(n+1)$ $\boxed{=>}$ $\dfrac{\pi}{2} = \lim\limits_{n \to +\infty} \dfrac{[2^n.n!]^4}{[(2n)!]^2.(2n+1)}$

Corollary: $\sqrt{\pi} = \Gamma(1/2) = \lim\limits_{n \to +\infty} \dfrac{(2^n.n!)^2}{(2n)!} . \dfrac{1}{\sqrt{n}}$

Wallis derived the formulas of K,L,M by the
following trigonometric reduction formula

$$\int \sin^n \theta.d\theta = -\frac{\cos\theta.\sin^{n-1}\theta}{n} + \frac{n-1}{n}.\int \sin^{n-2}\theta.d\theta \quad n>/2$$

In the interval of integration of K,L,M
- Only the even powers of $\sin\theta$ carry the π -

$\int_0^{\pi/2} \sin^{2n-1}\theta.d\theta = \dfrac{1}{2}.\dfrac{\Gamma(n).\Gamma(1/2)}{\Gamma(n+1/2)} = K$ $\bigg|$ $\Gamma(n) = \dfrac{n!}{n}$ $\bigg|$ $\Gamma(n+\frac{1}{2}) = \dfrac{1}{2^{2n}}.\dfrac{(2n)!}{2^n.n!}.\sqrt{\pi}$

$\int_0^{\pi/2} \sin^{2n}\theta.d\theta = \dfrac{1}{2}.\dfrac{\Gamma(n+1/2).\Gamma(1/2)}{\Gamma(n+1)} = L$ $\bigg|$ $\Gamma(n+1) = n!$ $\bigg|$ $\Gamma(\frac{1}{2}) = \sqrt{\pi}$

$\int_0^{\pi/2} \sin^{2n+1}\theta.d\theta = \dfrac{1}{2}.\dfrac{\Gamma(n+1).\Gamma(1/2)}{\Gamma(n+3/2)} = M$ $\bigg|$ $\Gamma(n+3/2) = (n+\frac{1}{2}).\Gamma(n+\frac{1}{2})$

$\boxed{\text{As } n \to +\infty \text{ and to the limit:}}$ $\dfrac{M}{K} = \dfrac{2n}{2n+1} = 1$

Since $0 < \sin\theta < 1$ for $0 < \theta < \pi/2$

- -

$\sin^{2n-1}\theta > \sin^{2n}\theta > \sin^{2n+1}\theta$ ie: $\boxed{0 < M < L < K}$

Inequalities of integrand functions implies inequalities
of integrals over the same interval of integration
(but not vice versa)

$\boxed{\text{As } n \to +\infty \text{ and to the limit:}}$ $L=M => (n+\frac{1}{2})^2.\Gamma^2(n+\frac{1}{2}) = \Gamma^2(n+1)$

The proof that: $\Gamma(1) = 0! = 1$ and that $\Gamma(1/2) = \sqrt{\pi}$

$$\Gamma(x) = \int_0^{+\infty} e^{-t} \cdot t^{x-1} \cdot dt = \lim_{n \to +\infty} \frac{n! \cdot n^x}{x.(x+1).(x+2). \ldots .(x+n)} \qquad \begin{array}{l} x \# 0 \\ x \# -1 \\ x \# -2 \\ \text{etc} \end{array}$$

$$\Gamma(1) = \lim_{n \to +\infty} \frac{n! \cdot n}{1.2.3 \ . \ (1+n)} = \lim_{n \to +\infty} \frac{n! \cdot n}{(n+1)!} = \lim_{n \to +\infty} \frac{n}{n+1} = 1 = 0!$$

- - - - - - - - - - -

$$\Gamma(\frac{1}{2}) = \lim_{n \to +\infty} \frac{\sqrt{n}}{\dfrac{1}{2}.\dfrac{\left[\dfrac{1}{2}+1\right]}{1}.\dfrac{\left[\dfrac{1}{2}+2\right]}{2}.\dfrac{\left[\dfrac{1}{2}+3\right]}{3}. \ldots .\dfrac{\left[\dfrac{1}{2}+n\right]}{n}}$$

$$\Gamma(\frac{1}{2}) = \lim_{n \to +\infty} \frac{\sqrt{n}}{\dfrac{1}{2}.\left[\dfrac{3}{2}\right].\left[\dfrac{5}{4}\right].\left[\dfrac{7}{6}\right].\left[\dfrac{9}{8}\right]. \ldots .\left[\dfrac{2n+1}{2n}\right]}$$

$$\Gamma(\frac{1}{2}) = \lim_{n \to +\infty} \frac{2 \ . \ 4 \ . \ 6 \ . \ 8 \ \ldots \ldots \ldots \ 2n}{3 \ . \ 5 \ . \ 7 \ . \ 9 \ . \ldots .(2n-1)} \ . \ \frac{2.\sqrt{n}}{2n+1}$$

$$\Gamma(\frac{1}{2}) = \lim_{n \to +\infty} \frac{(2^n .n!)^2}{(2n)!} \ . \ \frac{2.\sqrt{n}}{2n+1} = \lim_{n \to +\infty} \frac{(2^n .n!)^2}{(2n)!} \ . \ \frac{1}{\sqrt{n}} = \sqrt{\pi}$$

The Wallis product formula

The Wallis product formula:

As $n \to +\infty$ $(n+\frac{1}{2}).\Gamma^2(n+\frac{1}{2}) = \Gamma^2(n+1)$ $\boxed{=>}$ $\dfrac{\pi}{2} = \lim_{n \to +\infty} \dfrac{[2^n .n!]^4}{[(2n)!]^2 .(2n+1)}$

Just to get an idea:
- This topic is at very advanced level -

The solutions of Bessel's equation

$$x \cdot \frac{d}{dx}\left[x \cdot \frac{dy}{dx} \right] + (b^2 \cdot x^2 - p^2) \cdot y = 0 \quad \boxed{\text{Or}} \quad x^2 \cdot y'' + x \cdot y' + (b^2 \cdot x^2 - p^2) \cdot y = 0$$

$0 \backslash < x < +\infty$ p=any positive real number   b=constant#0
Or p=n (n=a positive integer or zero)

1st linearly independent solution	2nd linearly indep/nt solution
$Jp(bx) = \sum\limits_{k=0}^{+\infty} \dfrac{(-1)^k}{k!} \cdot \dfrac{\left[\dfrac{b.x}{2}\right]^{p+2k}}{\Gamma(p+k+1)}$	$J-p(bx) = \sum\limits_{k=0}^{+\infty} \dfrac{(-1)^k}{k!} \cdot \dfrac{\left[\dfrac{b.x}{2}\right]^{-p+2k}}{\Gamma(-p+k+1)}$

p=n $Jn(bx) = (-1)^n \cdot J-n(bx)$ **Linearly dependent**

Special forms:

$$J_{\frac{1}{2}}(bx) = \sqrt{\frac{2}{\pi.b.x}} \cdot \sin(b.x) \qquad\qquad J_{-\frac{1}{2}}(bx) = \sqrt{\frac{2}{\pi.b.x}} \cdot \cos(b.x)$$

The 2nd linearly independent solution for integers

$$Yn(bx) = \frac{2}{\pi} \cdot \left[\left[\ln\left(\frac{b.x}{2}\right) + \gamma\right] \cdot Jn(bx) - \frac{1}{2} \cdot \sum_{k=0}^{n-1} \frac{(n-k-1)!}{k!} \cdot \left[\frac{b.x}{2}\right]^{2k-n} \right. $$
$$\left. - \frac{1}{2} \cdot \sum_{k=0}^{+\infty} \frac{(-1)^k \cdot \left[\frac{b.x}{2}\right]^{2k+n}}{k! \cdot (k+n)!} \cdot [\varphi(k) + \varphi(k+n)] \right]$$

$\varphi(k) = \left[1 + \dfrac{1}{2} + .. + \dfrac{1}{k}\right]$ $\varphi(0)=0$ γ=Euler's constant=0.5777216
By convention

Summary of the properties of the solutions of the classical Bessel function and the basic integrals of those solutions

$$Bn(bx) = \lambda.Jn(bx) + \mu.Yn(bx) \mid \lambda,\mu \text{ constants}$$

Valid also for p instead of n
p=any positive real number

$$Bn_{-1}(bx) + Bn_{+1}(bx) = \frac{2.n}{b.x}.Bn(bx)$$

$$Bn(z) = \frac{2.(n-1)}{b.x}.Bn_{-1}(z) - Bn_{-2}(z)$$

$$z = (b.x)$$

$$Bn_{-1}(bx) - Bn_{+1}(bx) = \frac{2}{b}.\frac{d}{dx}Bn(bx)$$

$$\frac{d}{dx}Bn(z) = \frac{b}{2}.\left[Bn_{-1}(z) - Bn_{+1}(z)\right]$$

The basic integrals

1.
$$\int x^{n+1}.Bn(z).dx = \frac{1}{b}.x^{n+1}.Bn_{+1}(z) + C$$

2.
$$\int x^{1-n}.Bn(z).dx = -\frac{1}{b}.x^{1-n}.Bn_{-1}(z) + C$$

3.
$$\int Bn(bx).Bn(ax).x.dx = x.\frac{\begin{vmatrix} Bn(bx) & [d/dx\ Bn(bx)] \\ Bn(ax) & [d/dx\ Bn(ax)] \end{vmatrix}}{(b^2 - a^2)} + C$$

4.
$$\int Bn^2(bx).x.dx = \frac{1}{2.b^2}.\left[x^2.\left[\frac{d}{dx}Bn(bx)\right]^2 + \left[b^2.x^2 - n^2\right].Bn^2(bx)\right] + C$$

Formula most commonly used
for expansion in orthogonal functions

$$\int Bn^2(bx).x.dx = \frac{x^2}{2}.\left[\frac{1}{b^2}.\left[\frac{d}{dx}Bn(bx)\right]^2 + \left[1 - \frac{n^2}{b^2.x^2}\right].Bn^2(bx)\right] + C$$

Bessel's equation plays an important role in mathematical analysis

- Bessel's equation can be converted into Sturm-Liouville form necessary for **expansion in orthogonal functions** (Complete set of orthogonal functions)

  Another application of the delta function
  Orthogonality and Dirac go hand in hand

- Separation of three variables in 2nd order linear homogeneous differential equation's with partial derivatives (2oLHDEcPD) using polar coordinates.

Separation of the three variables

$$f[x(t), y(t)] = W(x,y).T(t) = U(x).V(y).T(t)$$
$$f[\ x[\rho(t), \theta(t)]\ ,\ y[\rho(t), \theta(t)]\ ] = M(\rho, \theta).T(t) = R(\rho).\Omega(\theta).T(t)$$

Suppose that: **Switching to polar coordinates**

$$\frac{d^2 f}{dx^2} + \frac{d^2 f}{dy^2} = \frac{1}{c^2}.\frac{d^2 f}{dt^2}$$

Or

$$\frac{d^2 f}{d\rho^2} + \frac{1}{\rho}.\frac{df}{d\rho} + \frac{1}{\rho^2}.\frac{d^2 f}{d\theta^2} = \frac{1}{c^2}.\frac{d^2 f}{dt^2}$$

Two equivalent forms - Both are separable

In general if the domain of the solution is bounded by straight line segments we prefer rectangular coordinates.

But if the domain is some segment or section of a circle , or circular annulus then the plane polar coordinates will be chosen.In other words the shape of the domain (ie the geometry of the domain) dictates the choise of the coordinates.

- Back to my method -

- All the properties of the trigonometric functions -
can be derived by the relation

$$e^{i.p} = [\cos p + i.\sin p]$$

$$e^{i.p} = e^{\xi} = 1 + \xi + \frac{\xi^2}{2!} + \frac{\xi^3}{3!} + \frac{\xi^4}{4!} + \frac{\xi^5}{5!} + \ldots$$ **Convergent infinite series for any real p (! Factorial)**

$$= [1 - p^2/2! + p^4/4! - ..] + i.[p - p^3/3! + p^5/5! - ...]$$

$$= \text{---------}\cos p\text{--------} + i. \text{---------}\sin p\text{---------}$$

Even powers-Even function   Odd powers-Odd function

Even: $f(x)=f(-x)$ **Odd:** $f(x)=-f(-x)$

$$e^{i.p}.e^{-i.p} = 1 = [\cos p + i.\sin p].[\cos p - i.\sin p] = [\cos^2 p + \sin^2 p]$$

$$e^{i.(p+q)} = [\cos(p+q) + i.\sin(p+q)]$$

$$i^2 = -1$$

$$e^{i.p}.e^{i.q} = [\cos p + i.\sin p].[\cos q + i.\sin q]$$

$\cos(p+q) = [\cos p.\cos q - \sin p.\sin q]$ **Real part**

$\sin(p+q) = [\sin p.\cos q + \cos p.\sin q]$ **Imaginary part**
without the i in front

.After many many years of self education in higher mathematics I reached a satisfactory level in various fields of this science.

Integral calculus is my favourite topic

. At one time in the past I tried (without success) to find a method to provide cook book formulas for the integrals of:

$$\int t^n .\cos(k.t).dt \text{ and } \int t^n .\sin(k.t).dt \text{ for high integer n. (n>2)}$$

.Along this direction I tried the integral $\int e^x .x^n .dx$ and bingo.

My sound knowlegde of the the gamma function $\Gamma(x)$ verified my general formula at the improper form of this integral.

$$\int_0^{+\infty} e^{-t} .t^{x-1} .dt = 2.\int_0^{+\infty} e^{-t^2} .t^{2x-1} .dt = \int_0^1 \left[\ln\left(\frac{1}{t}\right)\right]^{x-1} .dt = \Gamma(x)$$

$x\#0,-1,-2,etc \qquad x>0 \qquad x>0$

And: $\qquad\qquad n=0,1,2,3,\ldots,n \qquad n!=\Gamma(n+1)$

$$\int_0^{+\infty} e^{-s.t} .t^n .dt = 2.\int_0^{+\infty} e^{-s.t^2} .t^{2n+1} .dt = \int_0^1 t^{s-1} .\left[\ln\left(\frac{1}{t}\right)\right]^n .dt = \frac{\Gamma(n+1)}{s^{n+1}}$$

$s>0 \qquad\qquad s>0 \qquad\qquad s>0 \qquad\qquad s$

.Studying the differential equations I came across the method of undetermined coefficients from the reference 9.

$$r=(a+i.b)$$

I applied this method to the integral $\int e^{r.x} .x^n .dx$ and bingo.

Again the verification came from the improper forms of Laplace Transformation.

- The joy of discovery cannot be described -

J.K.Bredakis MD

References:

1. **Higher Mathematics for beginners:**
   ----
   by Ya.B.Zeldovich
   (Mir Publishers Moscow 1973)

2. **Calculus with analytic geometry:**
   ----
   by Harley Flanders and Justin J Price
   (Academic Press 1978)

3. **A brief course of higher mathematics:**
   ----
   by V.A.Kudryavtsev and B.P.Demidovich
   (Mir Publisher's Moscow 1980)

4. **Concice Encyclopedia of Mathematics:**
   ----
   by W.Gellert,H.Kustner,M Hellwich,H Kastner
   (Van Nostrand Reinhold Company New York and other cities 1977)

5. **Computational Mathematics:**
   ----
   by B.P Demitovich and I.A.Maron
   (Mir publishers Moscow 1976)

6. **Advanced calculus:**
   ----
   by Leopold Flatto
   (The Wiiliams and Wilkins Company - Baltimore 1982)

7. **Mathematics Handbook for Science and Engineering:**
   ----
   by: Royal Lennart Rade and Bertil Westegren
   Fifth edition - 2004
   Springer Verlag Publications Inc
   Berlin - Heidelberg - New York

8. **Mathematical methods for physicists and engineers:**
   ----
   by: Royal Eugene Collins - 2nd corrected edition
   Dover Publications Inc - Mineola New York - USA 1991

9. **Differential Equations:**

   A systems approach - by: Jack Goldberg - and Merle C.Potter
   Prentice Hall International Editions
   Upper Saddle River , NJ - USA - 1998

- **And a lot of personal work** -